PEDIGREE ANALYSIS: PROBLEM SOLVING

Darani Vasudevan

Preface to the book,

This book holds the tips that are required to solve the calculations related to pedigree analysis. This book would be useful to students, lecturers and to those who have interest in calculating inheritance of a trait. The book holds the pedigree analysis questions asked in CSIR UGC NET Life science examination. So this book will definitely form a hand in reference to CSIR NET, SET aspirants.

V.Darani M.Sc.,M.Phil.,(SET)

Chapter 1

What is a Pedigree?

A chart of a family showing the phenotype and relationships of the members of the family is called Pedigree.

The members of the family and their relationships in the pedigree chart are represented in the form of symbols.

The method used to determine the mode of inheritance or transmission of a genetic disorder or trait in a family is called Pedigree Analysis.

The symbols used in Pedigree chart are as follows.

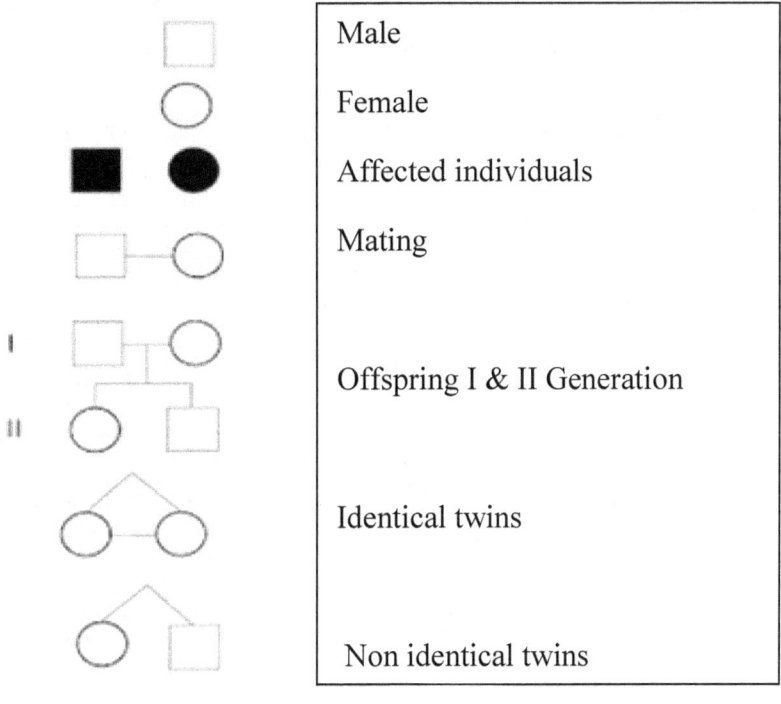

Chapter 2

Pedigree analysis- Calculations

In various competitive examinations, we frequently come across questions related to pedigree analysis. A pedigree chart will be given and we would be asked to find the type of inheritance or the probability of inheritance of a trait in that family. The Pedigree analysis is slight an advanced calculation related to Genetics. It will look difficult to solve for the first time, but once we are clear about the modes and patterns of inheritance we could solve any Pedigree questions within few seconds.

In CSIR NET Life Science examination conducted in India for the post of Junior Research Fellow and Assistant Professor, questions are frequently asked regarding Pedigree analysis. Thus we would not be able to solve it unless we have a clear idea and logic of such problems.

Pedigree Analysis is obviously an interesting topic. The following chapters deal with inheritance of a trait in animals and the tips and tricks for solving problems related to pedigree analysis.

Generally speaking, pedigree analysis problems are broadly grouped into two categories

- Genetics related – The questions related to this category generally ask the mode of inheritance whether it is autosomal or sex linked inheritance.
- Probability related – The questions based on this category would ask us to calculate the percentage of inheritance of a character or trait in the generation. The probability related questions require a little mathematical knowledge for calculation.

Chapter 3
Genetics Related Pedigree Analysis Calculations

What is Genetic inheritance?

We all know that each cell in the body possess 23 pairs of chromosomes, one half of each chromosome is inherited from our mother and the other half is inherited to us from our father. The chromosome thus has genes inherited from our parents and the dominant characters are expressed as phenotypes. This mode of inheritance of genes or characters from our parents is referred to as genetic inheritance.

Genetic Disorder

When a sudden change occurs in DNA, an alteration in the genetic inheritance occurs. This sudden change is referred by the term mutation. This alteration results in change in character or genetic disorder. This mutated gene may be passed onto a child by his or her parents. The genetic inheritance of a disorder can fall into any of the ways described below.

Autosomal Dominant inheritance

In this case the dominant allele will be mutated and is responsible for causing the disorder in the generation. In autosomal dominant inheritance, only one parent needs to carry the mutated gene. Examples of autosomal dominant inheritance are polycystic kidney disease, Huntington's disease, Hereditary Spherocytosis.

Autosomal Recessive inheritance

In this type, the recessive allele will be mutated and is responsible for causing the disorder. In autosomal recessive inheritance both the parents must have a copy of the mutated allele. In simple words, the parents must be carriers in this case. Examples of Autosomal recessive inheritance are Sickel cell anaemia, Cystic fibrosis, Tay–sachs disease.

X-linked Inheritance

The mutations in the sex chromosome or the X chromosome are referred as the X linked inheritance. The X linked inheritance are further classified into two types,

- ➢ X linked Dominant inheritance
- ➢ X linked Recessive inheritance

X linked Dominant Inheritance

It is less common when compared to X linked recessive mode of inheritance. The dominant allele on X chromosome is mutated in this case. This means that only one copy of the mutated allele is sufficient to cause the disorder. The females are more affected than males.
Examples: Rett syndrome, Vitamin D resistant rickets, Alport syndrome

X linked recessive Inheritance

In this mode of inheritance the mutation of a allele in the chromosome cause the phenotype to be expressed in males.
Examples: Haemophilia, Colour blindness

Y linked inheritance

It is also referred as holandric inheritance. The mutated allele responsible for causing the disorder will be located on the Y chromosome. This mode of inheritance is passed on only from father to son since the Y chromosomes are present only in males.

Examples: Hypertrichosis pinnae, Azoospermia, Retinitis pigmentosa.

Cytoplasmic (maternal) inheritance

Cytoplasmic inheritance is also referred to as Extranuclear inheritance. In this case the transmission of genes occurs outside the nucleus. It is commonly known to occur in cytoplasmic organelles such as mitochondria or chloroplast. In short it could be defined as the inheritance of characters or traits usually via maternal line and is controlled by an extrachromosomal element. Thus all the children will be affected if mother is affected.

Chapter 4

Pedigree Charts

Autosomal Dominant Inheritance

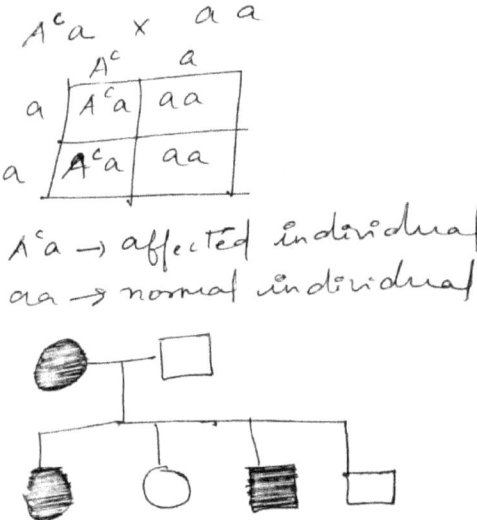

Autosomal Recessive Inheritance

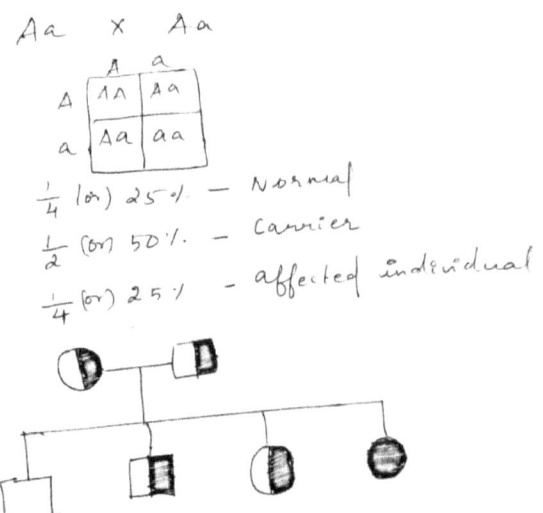

X- Linked Dominant Inheritance- Affected Mother

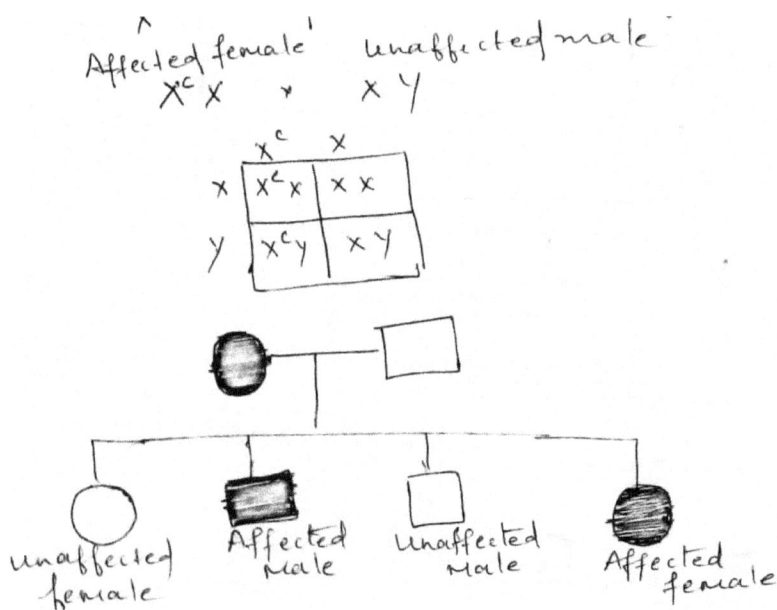

X- Linked Dominant Inheritance- Affected Father

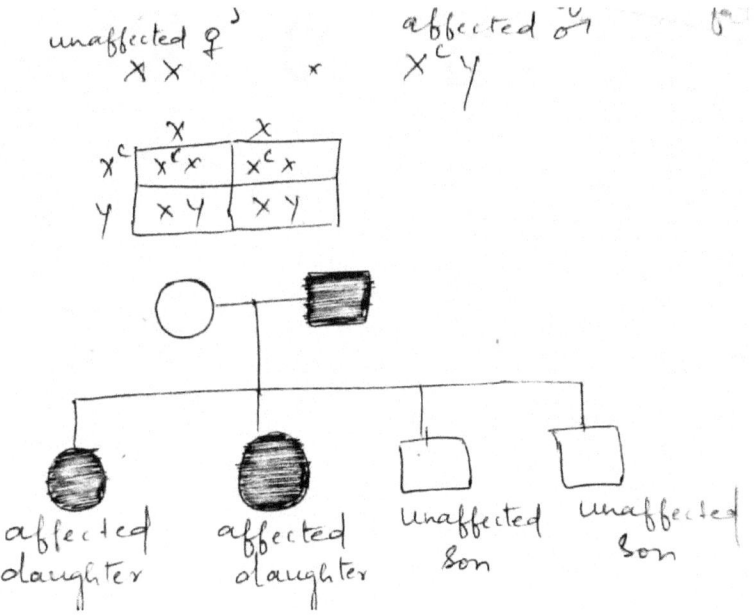

X- Linked Recessive Inheritance

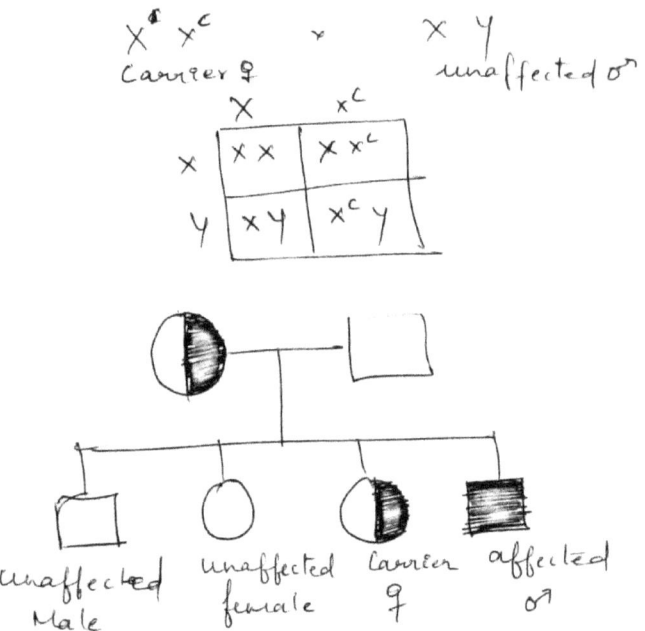

Chapter 5
Terms Related to inheritance

Unifactorial inheritance

Trait like blood type, eye colour, hair colour and taste each are controlled by simple pair of genes. Each trait may have different alleles.

Example: Hair colour brown or red

If father has brown/red, he has 50% chance of transferring brown to child and 50% chance of transferring red. If mother has brown/brown she has 100% chance of transferring brown.

Multifactorial inheritance

Some traits are determined by genes and environmental factors. For example, a child who has received tall genes determining the height from both the parents and grew taller than both the parents. On the other hand if the child has received dwarf genes from both the parents he would be the shortest in the family.

Polygenic traits

The traits are determined by the combined effects of many genes. For example, the size of all body parts determines the height together. Similarly eye, skin, hair is polygenic traits because they are determined by more than one allele at different location.

Intermediate expression

When there is incomplete dominance, blending can occur resulting in homozygous individuals. For example, pitch of a male's voice: Homozygous men will have the lowest and highest of this trait.

Co-Dominance

In co dominance, both the alleles are expressed in heterozygous individuals Example: AB blood group.

Incomplete penetration

Some genes penetrate incompletely, which means, they are not expressed unless certain environmental factors are present. For example, person may have gene for diabetes but doesn't show disease unless they are subjected to stress like overweight, not enough sleep etc.

Chapter 6
Points to remember while solving Genetics related Pedigree analysis Problems

General

- A normal individual cannot have alleles of the dominant trait.
- A normal individual can be a carrier of the recessive trait.

Autosomal Inheritance

- In an autosomal trait, the father transmits a trait to his son.
- In dominant traits, the phenotype appears every generation.
- If there is a skipping of generations then the trait is recessive.
- An individual with the recessive phenotype must be homozygous recessive.
- If the affected individual has both the parents normal, then we could conclude that the trait is recessive.
- In the case of dominant inheritance, the affected parents can have unaffected child.

X-linked inheritance

- If the trait is X linked, the males would be affected even by a single recessive allele.
- The trait is never passed on from father to son.
- The transmission of trait occurs from father to his daughters.
- If the normal parents have affected 'son', then the trait is recessive but we could not conclude whether it is autosomal or sex linked. If the trait is obtained from both the parents then it is autosomal. In the case of X linked the trait is obtained from his mother.

- In X linked inheritance if the mother is affected all her sons would be affected.
- In X linked recessive inheritance males are more affected than females.

Y- Linked inheritance

- There will be no affected or carrier females.
- The Y linked inheritance is mainly male to male transmission.

Chapter 7

Questions asked in CSIR NET related to Genetics related Pedigree analysis

The modes of inheritance explained in the pedigree below are:

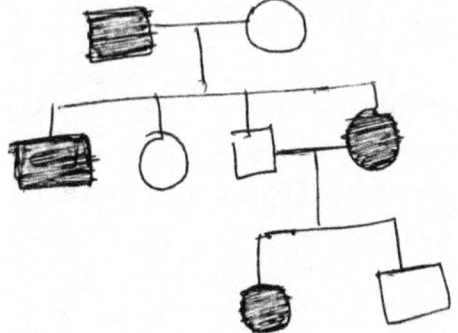

Answer: Autosomal recessive and Autosomal dominant

The below pedigree shows the inheritance of a rare allele. The allele is:

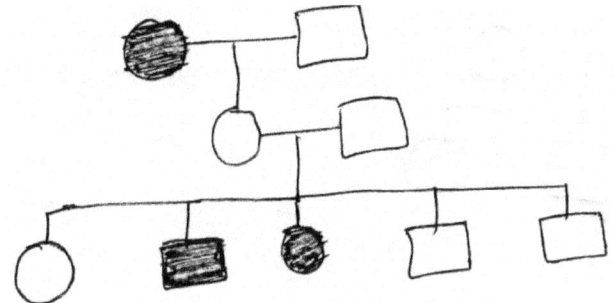

Answer: Dominant with incomplete penetrance.

Chapter 8
Tips for solving Probability related Pedigree analysis calculations

The ideas for solving the probability related pedigree analysis problems could be easily understood with the help of an example.

The pedigree shows the occurrence of an autosomal recessive trait, where the black squares have genotype aa. Calculate the probability of IV-1 will be either affected (aa) or a carrier (Aa).

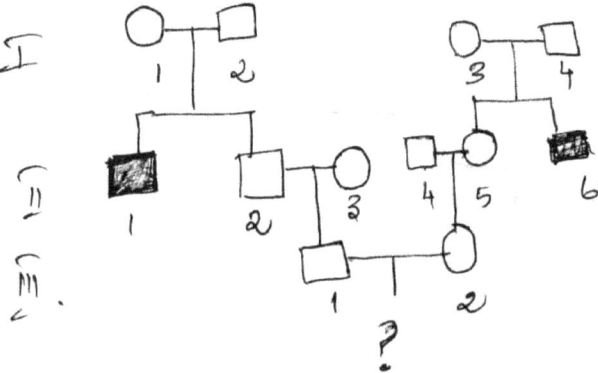

I-1 and I-2 are carriers because in the next generation affected individuals appeared. Hence their genotype could be devised as Aa.

The cross Aa × Aa is as follows

	A	a
A	AA	Aa
a	Aa	aa

AA = 1/4   aa = 1/4
Aa = 1/2
Probability of Aa = 2/3

The probability of normal individual is 1/4 or 25%

The probability of carrier individuals is 1/2 or 50%
The probability of affected individuals is 1/4 or 25%

From the above we could conclude the probability of II-2 to be a carrier is 2/3 (since, dominant allele is present in 3 of the four individuals, out of which 2 are carriers)
The individual II-3 belongs to a different family unaffected by the disorder and hence its genotype is AA.
A cross between II-2 and II-3 (Aa × AA) can be expressed as follows,

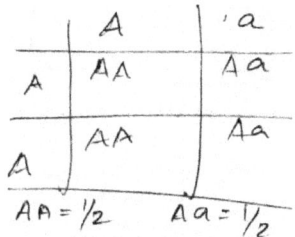

Thus the probability of appearance of carriers in the next generation is 1/2 or 50% and the probability of appearance of the normal individuals in the next generation is 1/2 or 50%

Therefore, the probability of III -1 to be heterozygous is 2/3×1/2=1/3 (The individuals probability must be multiplied with the carrier percentage of the previous generation).

The same steps are applied for finding the carrier probability of III-2. The carrier probability of III-2 is thus 1/3.

A cross between III-1 and III-2 (Aa × Aa) is as follows,

	A	a
A	AA	Aa
a	Aa	aa

AA = 1/4   aa = 1/4
Aa = 1/2
Probability of Aa = 2/3

From the table, the probability of appearance of carriers in the next generation is 1/4 and the probability of appearance of carriers in the next generation is 1/2.

Probability of IV-1 to possess the affected trait

The probability of IV-1 to have affected trait is 1/3×1/3×1/4=1/36. This means that, 1 in 36 individuals are affected with the trait.

Probability of IV-1 to possess the affected trait

The probability of IV 1 to be carrier is:
If one of the parent is homozygous dominant and the other is heterozygous:
AA× Aa
The probability of III 1 to be Homozygous dominant is 2/3 (if 1/3 of the individuals are heterozygous the remaining 2/3 will be homozygous dominant).
2/3×1/3×1/2= 2/18=1/9
If both the parents are heterozygous the probability of IV1 to be carrier is
Aa×Aa
1/3×1/3×1/2=1/18.

Chapter 9
More examples for Better Understanding

Dev's grandaunt (grandfather's younger sister) had Tay Sachs due to mutated TS gene. That's the only case in the grandfather's family. What is the probability of Dev's being a carrier of the mutated allele.

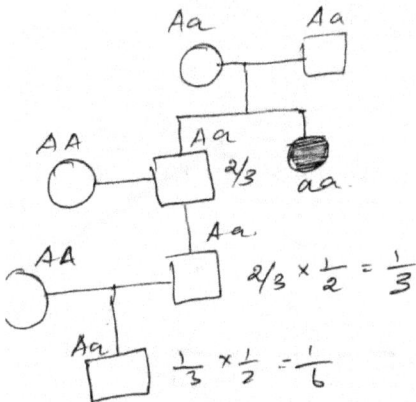

A married couple went to see a genetic counsellor because each had a sibling with sickle cell anaemia. SCA is a recessive disorder and neither member of the couple nor any of their four parents are affected. What are the chances that their child will be affected with SCA?

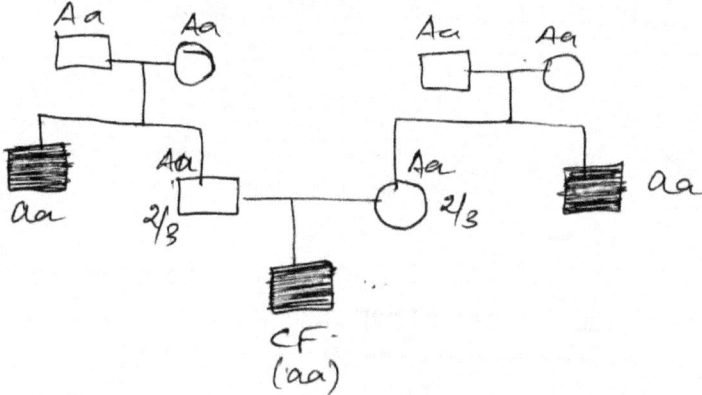

The probability of II2 and II 3 to be heterozygous is 2/3 each.
A cross between them is as follows:

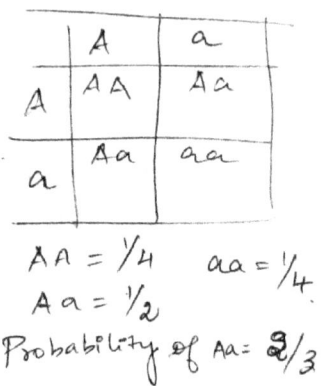

$AA = 1/4 \qquad aa = 1/4$
$Aa = 1/2$
Probability of Aa = 2/3

The probability of III 1 to be affected with SCA is 2/3×2/3×1/4 = 4/36 = 1/9.

From the given autosomal recessive trait based chart find the probability of the expression of the trait by V 1

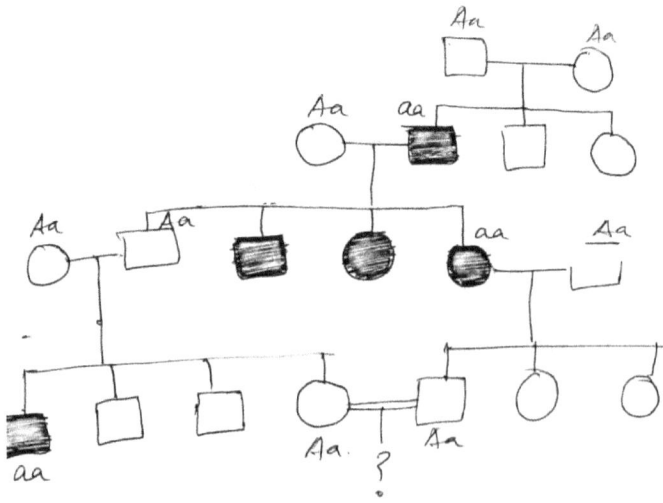

In the first generation both the parents must be heterozygous (Aa) since the second generation have affected individuals.
II 2 is affected and hence its genotype would be aa and II 1 must be heterozygous since the third generation produced by a cross between II 1 and II 2 carries affected individuals. So the genotype of II 1 is Aa.
With the same logic we could assume the genotype of III 1 and III 2 are Aa and Aa respectively.

The probability of IV 4 to be heterozygous is 2/3
The genotype of III 5 is aa since it is an affected individual.
Since the fourth generation doesn't hold any affected individual the genotype of III 6 is AA.
The cross between AA× aa

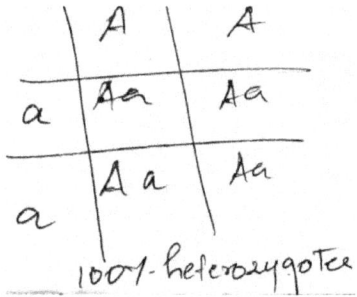

Thus the probability of IV 5 to be heterozygous is 1/1.
From the cross between Aa × Aa, we could say the probability of V 1 to be affected with the trait is 1/4.
Therefore the overall probability would be 2/3 × 1/1 × 1/4 = 2/12 =1/6.

Karan and Arjun each have a sibling with cystic fibrosis. Neither Karan or Arjun nor any of their parents has the disease and none of them has been tested to reveal cystic fibrosis trait. Based on this incomplete information, calculate the probability that if this couple should have another child, the child will have cystic fibrosis.

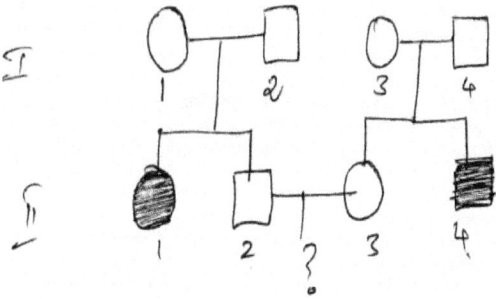

It could be easily calculated that the probability of III 1 to be affected with cystic fibrosis is 2/3 × 2/3 × 1//4 = 4/36 = 1/9.

Chapter 10
Questions related to Probability based Pedigree analysis in CSIR NET

The inheritance pattern of a common trait which shows complete penetrance is shown below:

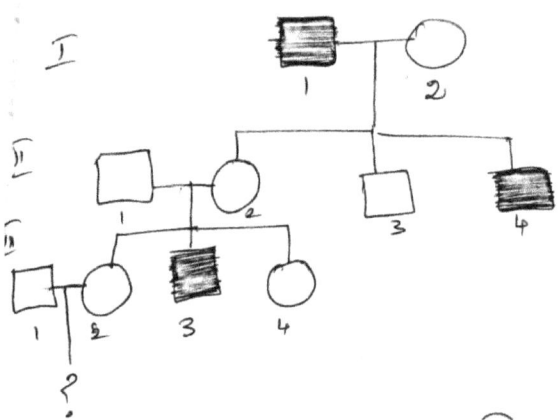

Based on the above pedigree, fill in the blanks given below:
The trait is -------. The probability that a child from the marriage of individual III 1 and III 2 will show the trait is ----- considering that the individual III 1 is a carrier of the trait.

Solution:
III 1 the probability of being carrier is 1/1, the probability of III 2 being heterozygous is 1/3. From this the probability of IV 1 to be heterozygous is 1/1 × 1/3 × 1/2 = 1/6

Affected individuals from the pedigree given below are suffering from albinism, an autosomal recessive disease. Identify the confirmed carrier individuals in their pedigree assuming that the members coming from outside the family are homozygous for the dominant allele.

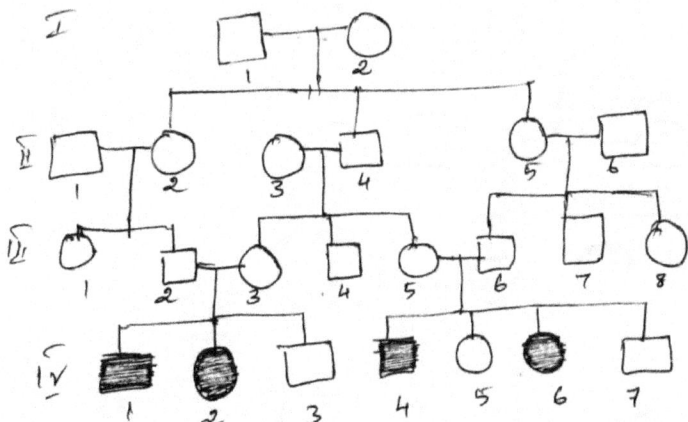

Answer: III 2, III 3, III 5, III 6, II 2, II 4, II 5 and I 2

The following pedigree shows the inheritance pattern of a rare recessive disorder with complete penetrance.

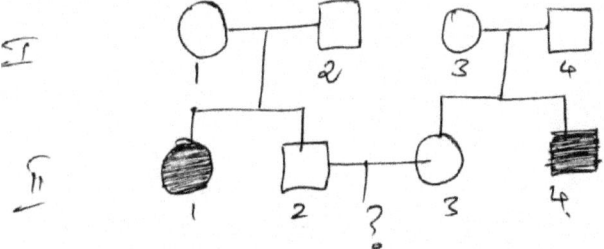

A child from marriage between individuals II 2 and II 3 will show the disorder only if the parents carry the recessive allele. What is the probability that the child will show the disorder?

Answer:

1/9, and the probability of the parents to carry the recessive allele is 2/3.